Valleys of Europe:

7. Norway (fjords)
8. Iceland (fjords)
9. Scotland (lochs)
10. Germany (Rhine valley)
11. France (Rhone Valley)
12. Italy (Po valley)
13. Alps (many passes)

...eys of Asia:

14. The main Himalayan rivers such as the Brahmaputra, the Ganges, and the Indus all cut deep valleys in the mountains.

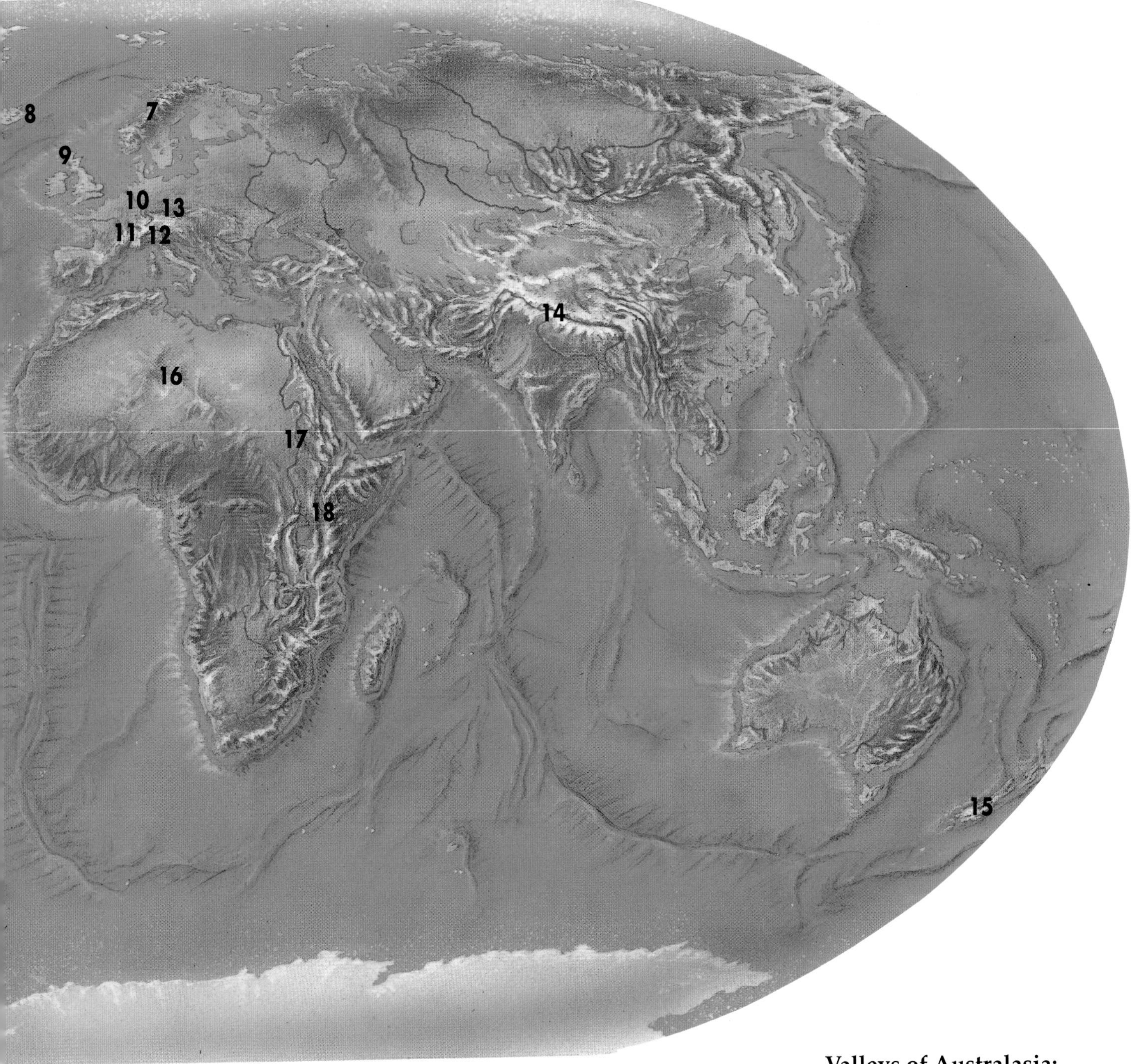

Valleys of Australasia:

15. New Zealand (sounds)

Valleys of Africa:

16. North Africa (wadis – old desert valleys)
17. Nile valley
18. East African Rift Valley

FACTS ABOUT VALLEYS

Rivers and their tributaries often cut many valleys as they flow from their sources to the sea. It is therefore unusual for any single valley to be very long. The largest steep valley (gorge) in the world is the Grand Canyon, Arizona, USA. It is 217 miles long and 5300 feet deep and up to 13 miles wide. (It is described more fully in the Landshapes book *Canyon*.)

The deepest valley in a mountainous region is in Nepal where the Kali River flows in a valley 18,000 feet deep. The longest river valleys in the world are in the Himalayas of Asia. The Brahmaputra River flows in a steep-sided valley flanked by mountains for over 800 miles. The Fraser River valley (that includes Banff and Jasper National Parks in Alberta, Canada) is 600 miles long. The Rio Grande valley is also over 400 miles long upstream of El Paso, Texas, USA.

When the Earth's crust pulls apart, long rift valleys are formed. The longest rift valley is the East African Rift Valley system which stretches from one end of the Jordan Valley in Lebanon through the Red Sea and East Africa to Mozambique, a total distance of over 2400 miles. Other important rift valleys include the valleys of the rivers Rhone and Saone in France which have a combined length of 400 miles and the River Rhine in Germany between Basel and Cologne, also about 400 miles.

The floor of Death Valley in California, USA, is the lowest place in the western hemisphere at 282 ft *below* sea level. The highest large valleys in the world often make passes over mountains. The highest valley pass in the world capable of taking road traffic is the 700 mile pass between Tibet and China. It includes a pass at nearly 18,500 ft.

The longest valley (fjord) connected to the sea is the Nordvest arm of the Scoresby Sund in Greenland, 188 miles.

Grolier Educational Corporation
SHERMAN TURNPIKE, DANBURY, CONNECTICUT 06816

VALLEY

Author
Brian Knapp, BSc, PhD
Art Director
Duncan McCrae, BSc
Editor
Rita Owen
Illustrator
David Hardy
Print consultants
Landmark Production Consultants Ltd
Printed and bound in Hong Kong
Designed and produced by
EARTHSCAPE EDITIONS

First published in the USA in 1993 by
GROLIER EDUCATIONAL CORPORATION,
Sherman Turnpike, Danbury, CT 06816

Library of Congress #92–072045

Cataloging information may be obtained
directly from Grolier Educational Corporation

Title ISBN 0–7172–7179–X

Set ISBN 0–7172–7176–5

Acknowledgements. The publishers would like to thank the following: Leighton Park School, Martin Morris and Redlands County Primary School.

Picture credits. All photographs from the Earthscape Editions photographic library except the following (t=top, b=bottom, l=left, r=right): ZEFA 8/9, 15t, 28l, 29, 34/35, 35r.
Cover picture: Cortina, Dolomites, Italy.

In this book you will find some words that have been shown in **bold** type. There is a full explanation of each of these words on page 36.

On many pages you will find experiments that you might like to try for yourself. They have been put in a blue box like this.

In this book mi means miles and ft means feet.

These people appear on a number of pages to help you to know the size of some landshapes.

CONTENTS

Introduction

A valley is a long trench between hills or mountains. Sometimes it forms a deep, narrow slot in the land and is called a gorge. If it is a little wider it is called a canyon. But most valleys will have moderate slopes and a wide flat floor across which a river makes its way in winding curves.

If the valley has sloping sides and looks like a V in **profile**, the valley has probably been cut by a river. Most of the world's valleys are river valleys. Other valleys have a more U shape – these have largely been cut by **glaciers**. Yet others are deep and very wide with flat floors; these have been made by forces deep within the Earth. They are called **rift** valleys.

Valleys change depending on how much **energy** rivers or ice may have to make them deeper or wider. If there is no shortage of energy then valleys become deeper; if there is less energy, then valleys may actually begin to fill in.

In this book you will find out about the many types of valleys in the world, how they were formed and how they are still changing. Enjoy the world of valleys by turning to a page of your choice.

The picture on this page is of mist-shrouded valleys in the Appalachian region of the USA.

Take care in valleys

Valleys can be spectacular places and you are sure to want to visit them. However, do take special care when visiting gorges and other steep-sided valleys and go with an experienced adult. Remember that valley sides may have loose rocks that can fall on to you, and that people have fallen into rivers and been carried away by swift currents.

Chapter 1
River valleys

What is a river valley?

If you walk up a valley side from any river bank you will eventually reach the place called the **divide** that marks the boundary between the valley you are in and its neighbor. Take one more step across this summit and you will be entering the neighboring river valley.

Every river valley is ringed by a divide. There are small divides that separate tiny streams within a large valley and there are large divides that separate groups of valleys.

Here is how they fit together.

The divide is the line that separates the water draining to neighboring rivers.

Each river cuts into its bed to make the basin deeper (see page 12).

Each valley side is an active slope (see page 14).

How does a river valley form?

A river valley is a combination of two processes: the downcutting by rivers, and the widening of slopes by **landslides, mudflows** and **soil creep**.

The pattern of valleys shown on this page has been made by many streams. Each valley fits neatly against the next just like pieces in a jigsaw. In this way the whole area is drained.

A region of many valleys and rivers that feed into one another is called a basin.

This is a **tributary** valley. Its stream drains the edge of the basin.
These steep-sided valleys are most commonly found in uplands (see page 18).
SCALE
Valleys change shape and slope over time (see page 16).
This is the main valley, fed by all the tributaries. Its floor is wide (see page 20).

The work of rivers

Rivers cut into the land by scouring their beds with the stones they carry during times of high flow. They also carry away material that falls into the channel when bank sides collapse.

Carrying material uses up a lot of the energy in a river, so the more material the river carries the less energy it has for cutting down.

In places where little material moves down slopes, nearly all the energy of a river is available to cut deeper and deeper into the bed, eventually creating a gorge.

Cutting down

In some parts of the world, such as in the US canyonlands shown below, you can clearly see what happens when river erosion is the main activity.

The river is armed with millions of tiny sand grains that scrape and scour the bottom of the river channel as they are carried along by the fast-flowing water. As you can see, rivers alone cannot make valleys, they simply cut slots or trenches.

For more information on gorges in dry places see the book Canyon *in the Landshapes set.*

A tributary enters the main river in its own gorge.

As little rain falls on this tableland, the break down of rock into small pieces is very slow.

The river works by cutting a deep slot. In this case it has vertical sides.

This road gives you an idea of the size and scale of the gorge.

River gorges
Gorges occur in rivers when the land rises across their paths or when a band of hard rock occurs.

A major river cutting a gorge. Notice how much silt is in the water: it has turned the river a brown color.

The gorge has nearly vertical walls.

Each band of rock is clearly seen.

Stream gorges
Gorges are common in mountains where the steep slopes give even streams enormous energy for cutting into their beds.

This road gives you an idea of the size and scale of the gorge.

The river fills the bottom of the gorge.

For more information about rivers see the book River *in the Landshapes set.*

How valley sides widen

In many valleys, soil and rocks from the valley sides tumble, slip and slide down into the river below. Carrying this debris uses up some of the energy of the river and gives a chance for the sides to widen and become less steep.

There are easy clues to help you see where soil and rocks are on the move, and how often they may slip and slide.

Scree

Rocks on steep slopes regularly tumble to the bottom because, once loosened, they cannot stand up at a very steep angle. You can see how a scree works by using a pile of dry sand. You will soon find that, no matter how hard you try, the sand will collapse when you try to build it too steeply.

Rocks can only tumble when they break loose. Frost can crack open rocks, making the rock splinters that will fall to give **scree**. These bare rocks are the source of the boulders that make up the scree near the river.

The trees give you an idea of the size and scale of the scree.

This scree material all rests at the same angle. As no trees have grown you can tell that the scree must be frequently moving.

Avalanches
In mountainous areas snow often builds into thick drifts during the winter. If the drifts lie at angles between 20 and 40 degrees they frequently collapse and make avalanches. Avalanches are very effective at tearing rock and soil from a valley side and transporting it to the valley floor.

Landslides
Where slopes are not steep enough for loosened rock to fall, the rocks break down into soil. This makes the material for a landslide.

A landslide occurs when a patch of soil gets very wet and loses its grip on the rock below.

Landslides leave big horseshoe-shaped scars, so they are easy to spot.

The cycle of valleys

There are many shapes and sizes of river valleys, but they all fit into a simple pattern of change.

To make it easier to understand the way the changes occur as a valley grows, is worn down and made wider, the name for each stage is often likened to the life cycle of people – youth, maturity and old age.

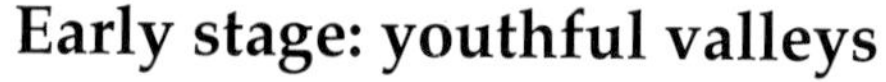

Early stage: youthful valleys

In mountain or very hilly areas rivers or glaciers can cut swiftly into the rock, carving steep-sided and narrow valleys. The valleys are still quite small and so they are sometimes called 'youthful' valleys.

Middle stage: mature valleys

In the more gentle hill regions, rivers cut sideways into the valley floors as well as down. These valleys have flat floors, or flood plains, with rivers curving across them. This is called the 'mature stage' of a valley, and changes happen here more slowly than in mountains.

Trees have been able to grow on the slopes. This only happens when rivers cut down less quickly.

This river still has 'white water' – places where the water is disturbed and small waves break. However, the river is much broader and less swift than in the early stage.

Final stage: old age valleys

Valleys near the sea sometimes have very broad valleys, with gentle hills far from where the river flows. These landscapes change most slowly of all. This is the 'old-age' stage of a valley.

The land is soil covered because of the gentle slopes.

There are no signs of fast-flowing water here. The river can carry only fine materials such as soil.

Upland valleys

The shape of a valley is a delicate balance between how fast a river cuts down into its bed against how fast soil and rocks can tumble off the slopes and make it wider.

In upland areas streams are usually so powerful that the valleys develop narrow shapes, rather like a capital letter V.

Soil and rock can only tumble to the water if the river keeps cutting down. This is the reason youthful valleys always have steep slopes.

This fast flowing river with rapids and much swirling water is cutting quickly into its bed.

Interlocking spurs

Even in uplands, where the slopes are very steep, rivers never follow a straight path towards the sea. Instead, the river sweeps across the land in a series of curves.

The curves made by the river give a winding valley and they prevent you from seeing along its whole length. The jutting out curves are called spurs and where there are many of them jutting out alternately from right and left, they are called interlocking spurs. They are a sign that the river is still in its youthful stage.

Note: one way to tell if an upland valley was carved by a river is to look for intelocking spurs. If, however, you can see straight along the valley, then it was cut by a glacier, as shown on page 22.

Each of these jutting ridges of land is called an interlocking spur.

The curves of the river and the valley are the same. Use the curves of the road to pick out the spurs.

From upland to lowland

This river shows the way upland rivers slowly change to those of lowland rivers. Notice how the river has widened the valley floor a little and that some of the material carried by the river has been dropped to make a sandy bed.

This means that the river is not cutting down as fast as it used to in the past.

Lowland valleys

The key sign of a change from a youthful, upland valley to a mature, lowland valley is when the river begins to curve around in the valley bottom. The flat floor is called a **flood plain**.

Flood plains show that the rivers cannot immediately carry away all the material that is being washed down from the uplands. So the rivers keep what they cannot carry in a kind of store room on their valley floors. Rivers can now curve, or **meander**, over this material.

This is the outside bend of a meander. As the river sweeps around this curve it cuts into the bank.

This is the edge of the flood plain.

This is the inside bend of a meander. It is the place where a river drops the silt and sand it cannot carry.

The banks are made from silt and sand laid down by the river. They are not made of rock.

Past signs of the way the river moves across its plain are seen in small curved lakes, called **oxbow lakes**.

The sides of the valley are quite gentle and covered in forest.

This is the edge of the flood plain. In this case it is made easy to see because farmers have only ploughed the flood plain.

Tidy meanders

Here the water in the river swings about, cutting into the outside bend of the meander, dropping material in the slack water of the inside bend.

By cutting and filling, the river is able to change its course and also move the stored sand, silt and clay a little farther downstream.

Valley's end

As rivers get close to the sea many stop cutting down and consequently valleys become very wide and shallow.

Some valleys are many miles wide at the coast and it is often very difficult to tell where one valley stops and another begins. In this picture you can see just a tiny part of the valley, that part of a coastal plain on which this river is laying down thick spreads of silt.

Chapter 2
Ice-cut valleys

U-shaped valleys

Rivers of ice – glaciers – are some of nature's fiercest tools. Glaciers can scour away as much rock from a mountain in a few thousand years as would take rivers millions of years.

Glaciers cannot flow around all the twists and turns of a river valley, so during an **Ice Age**, when glaciers fill these valleys, new shapes are created. Usually these are deep, straight trenches, much like a capital letter U. This is why they are called U-shaped valleys.

What glaciers change
The valley in the picture below suffered much erosion during the Ice Age. The floor has been deepened and widened, and the course is much straighter – you can easily see for many miles along its path.

The upper slopes of the valley are much more rugged than they would once have been. This is a sign that frost was an important factor in the past.

For more information on the work of glaciers see Glacier *in the Landshapes set.*

The world of ice

Although glaciers move slowly they are extremely powerful, taking along with them small flakes of rock that scour the valley floor and quickly cut a trench.

This picture shows a part of Greenland where there are many active glaciers.

Notice how these glaciers fill the bottom of each valley. Scouring can therefore take place across the entire valley floor.

Main glacier.

The glaciers have gouged straight paths where once there were interlocking spurs of river valleys.

Smaller tributary glaciers.

Hanging valleys

Unlike rivers, where tributaries feed neatly into main rivers, large and small glaciers cut their valleys almost independently.

Large valley glaciers are immensely powerful and they quickly cut deep U-shaped trenches. On the other hand, small glaciers in tributary valleys cut much more slowly and they are quickly outpaced by the main glacier and left 'hanging' above the main valley floor.

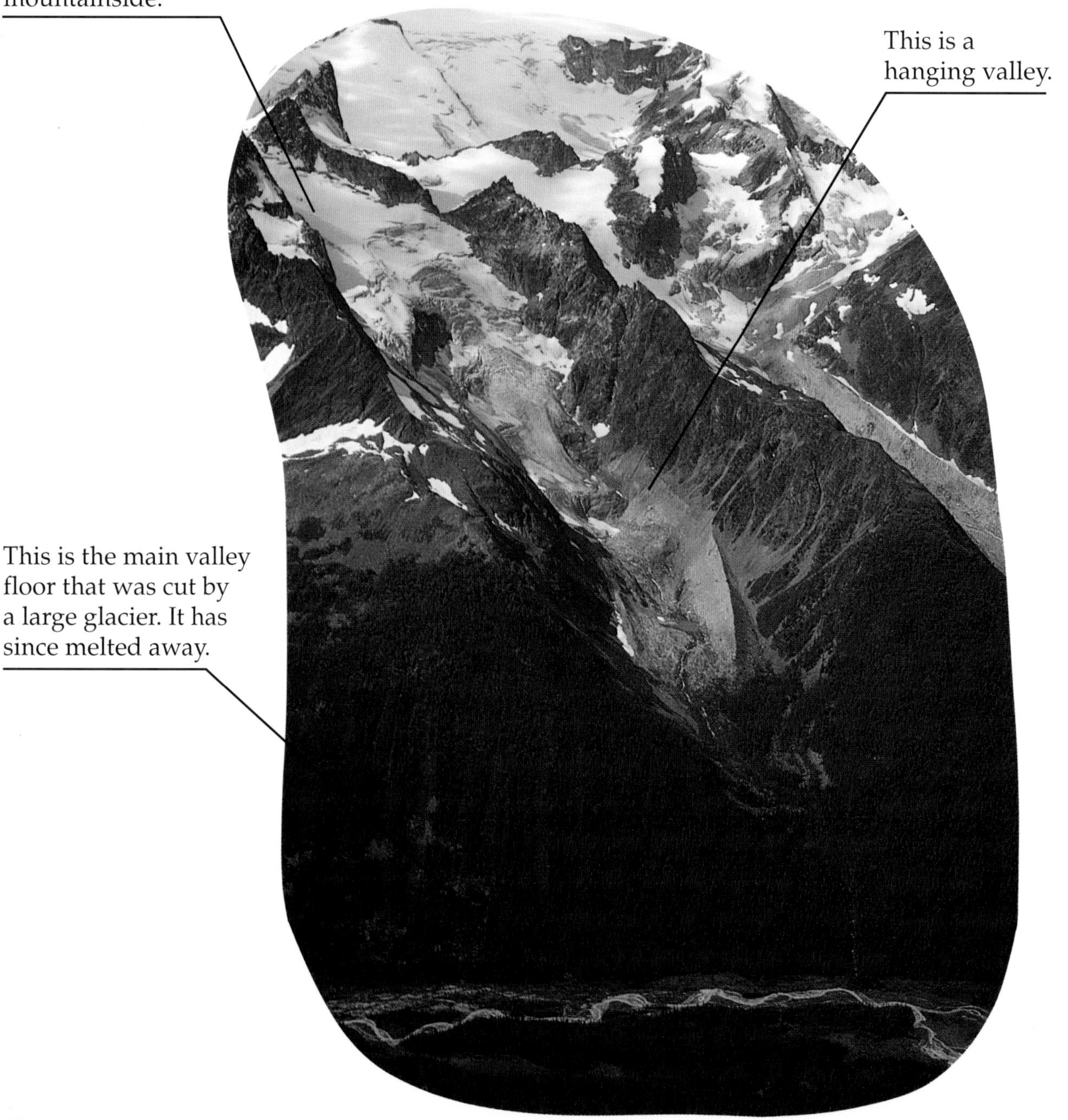

The hanging valleys in this region of Canada still have ice in them because they are very high on the mountainside.

This is a hanging valley.

This is the main valley floor that was cut by a large glacier. It has since melted away.

This landscape in Yosemite National Park, USA, has two hanging valleys. Each one has a waterfall to mark it. The picture below clearly shows both the Nevada (upper) and Vernal (lower) falls.

Tell-tale waterfalls

Valleys that have been left hanging are not easy to spot when they are still filled with ice.

When the ice has gone, rivers flow once again at the bottom of the valleys. Now it is much easier to spot the hanging valley as rivers often make spectacular waterfalls as they join the main valley.

Many of the world's tallest waterfalls are found in hanging valleys.

Passes

The world's highest mountains would be impossible to cross if it were not for the deep valleys that are used as passes.

The world's great passes are, however, unusual valleys because they have cut right across mountain ranges. This could only have been done by ice.

In many places passes are a means of getting from one place to another more easily. This picture shows a new motorway under construction through a pass in northern Italy.

Cutting across mountains
This is a pass high up in the Swiss Alps. The photograph was taken from one side of the summit of the pass and you can see right across the summit to the mountains that border the valley beyond.

How passes are formed

Large glaciers are far more powerful than small ones. Where a small glacier flows into a large one it may therefore not have the strength to force its ice into the main flow and it may become blocked. As a result the ice accumulates inside the smaller valley, getting ever thicker.

Imagine what happens as the ice completely fills the valley. It will begin to spill across its valley walls and into a neighbouring valley.

As the ice spills it scours the rock just as it does at the bottom of its valley. In this way a new trench is soon cut, this time between two valleys. When the ice melts away this trench will form a pass.

1. In the picture below you can see ice flowing inside the valleys.

2. In the picture below you can see thicker ice. It has already filled the valleys and has made a thick sheet called an ice cap. Passes are being cut underneath this ice cap.

Fjords

Where mountains are close to the coast, the land is often cut with deep valleys that have been flooded by the sea.

The Norwegians call these coastal valleys fjords. In Canada and New Zealand they are often called 'channels' or 'sounds'.

Although they appear to be long arms of the sea projecting inland, they were actually formed when the sea was many hundreds of miles away and the valleys filled with ice.

Norway's fjords

Norway has more fjords than any other country. Added together, the fjords make the Norwegian coast over 12,000 miles long: enough to stretch half way round the world.

Below is a picture of the Sogne Fjord, the longest fjord in Norway. The ocean liner shows you just how deep the water is even at the head of the fjord.

Submerged valleys

This picture shows a fjord in Greenland which still has a glacier in part of the valley. During the Ice Age, the sea level was much lower and the glacier would have flowed for hundreds of miles farther along the valley.

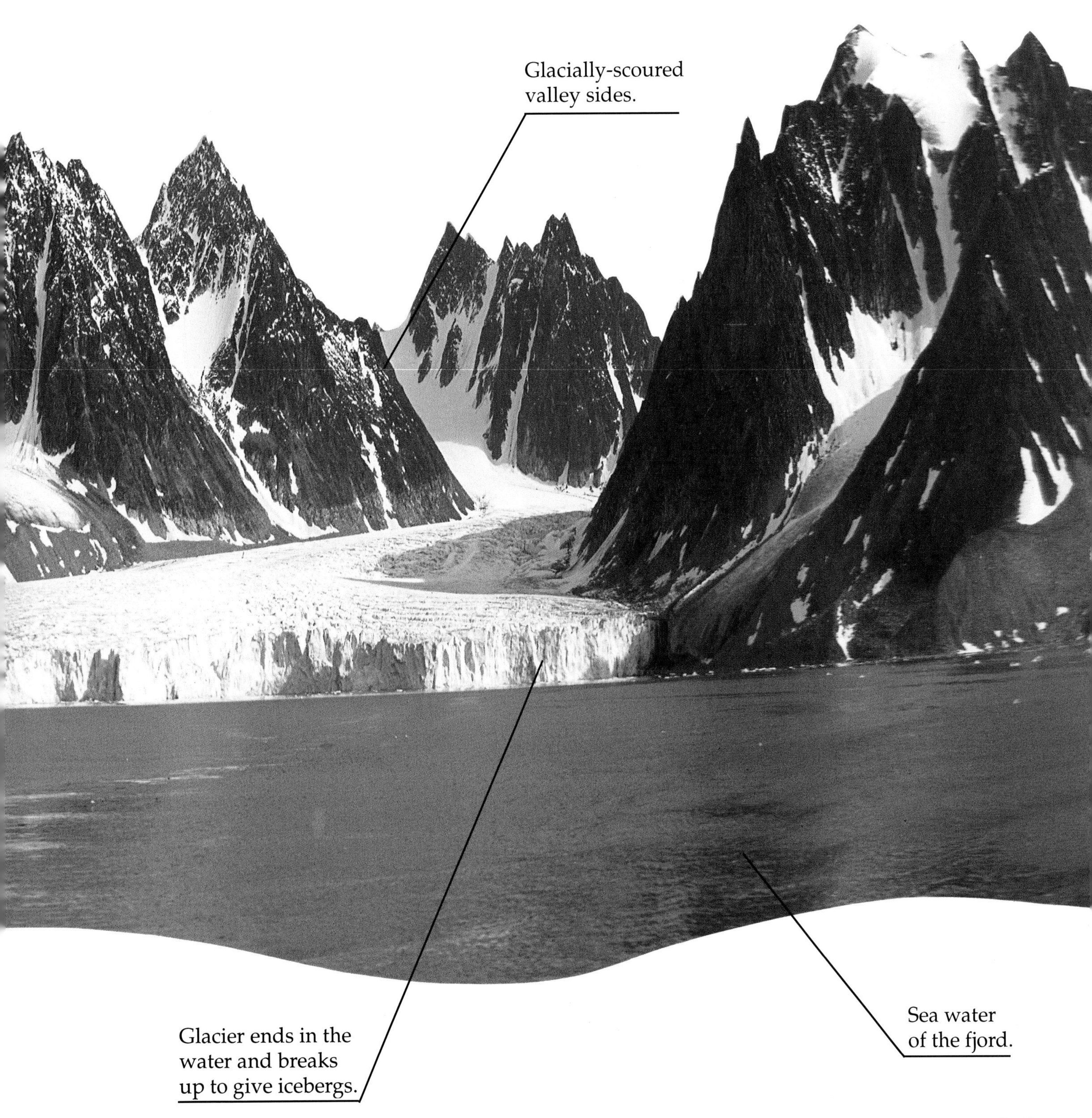

Chapter 3

Rift valleys

Fault-guided valleys

The world's largest valleys are rift valleys. They were not formed by rivers or tongues of ice, but by forces deep within the Earth that split, or faulted, the rocks.

A rift valley forms when two pieces of the Earth's surface **crust** begin to pull apart. The crust moves in occasional small jerks, so it may take millions of years for large rift valleys to form. However, when each jerk happens, people over a large area know about it – because each jerk makes an earthquake.

Rift valleys often occur in places where the crust is being stretched by forces within the Earth. Because the crust is brittle it begins to split apart.

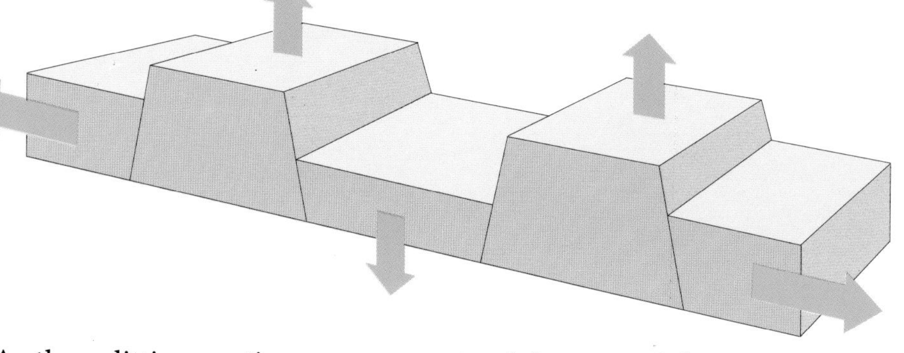

As the splitting continues, some parts of the crust sink down and make rift valleys. The rest of the land remains at the same height and now makes mountain ranges.

Deep canyons have formed in the sides of the valley. They have been formed by fast-flowing streams cascading down the steep valley walls.

Death Valley

Death Valley is a medium-sized rift valley in an area of south west USA that is criss-crossed by rift valleys. Its sides run straight for 140 mi, rising to a peak that is 11,000 ft above sea level, while the valley floor is 282 ft below sea level.

Although the floor of Death Valley is still sinking, the material worn from the valley sides is spread over the valley floor and this makes the true depth impossible to see. In fact the rock floor of the valley is at least 3000 ft below sea level!

Chapter 4
Valleys of the world

Yosemite

Many valleys have had a very complex history. Often several processes have been responsible for the shape we see today. Unravelling the history of a valley can be fascinating using the many clues that the landscape provides.

Yosemite Valley in California, for example, was formed by rivers, guided by faults, changed by glaciers, and is once again being changed by rivers. All of these activities have made this valley so spectacular that it has been protected as a National Park.

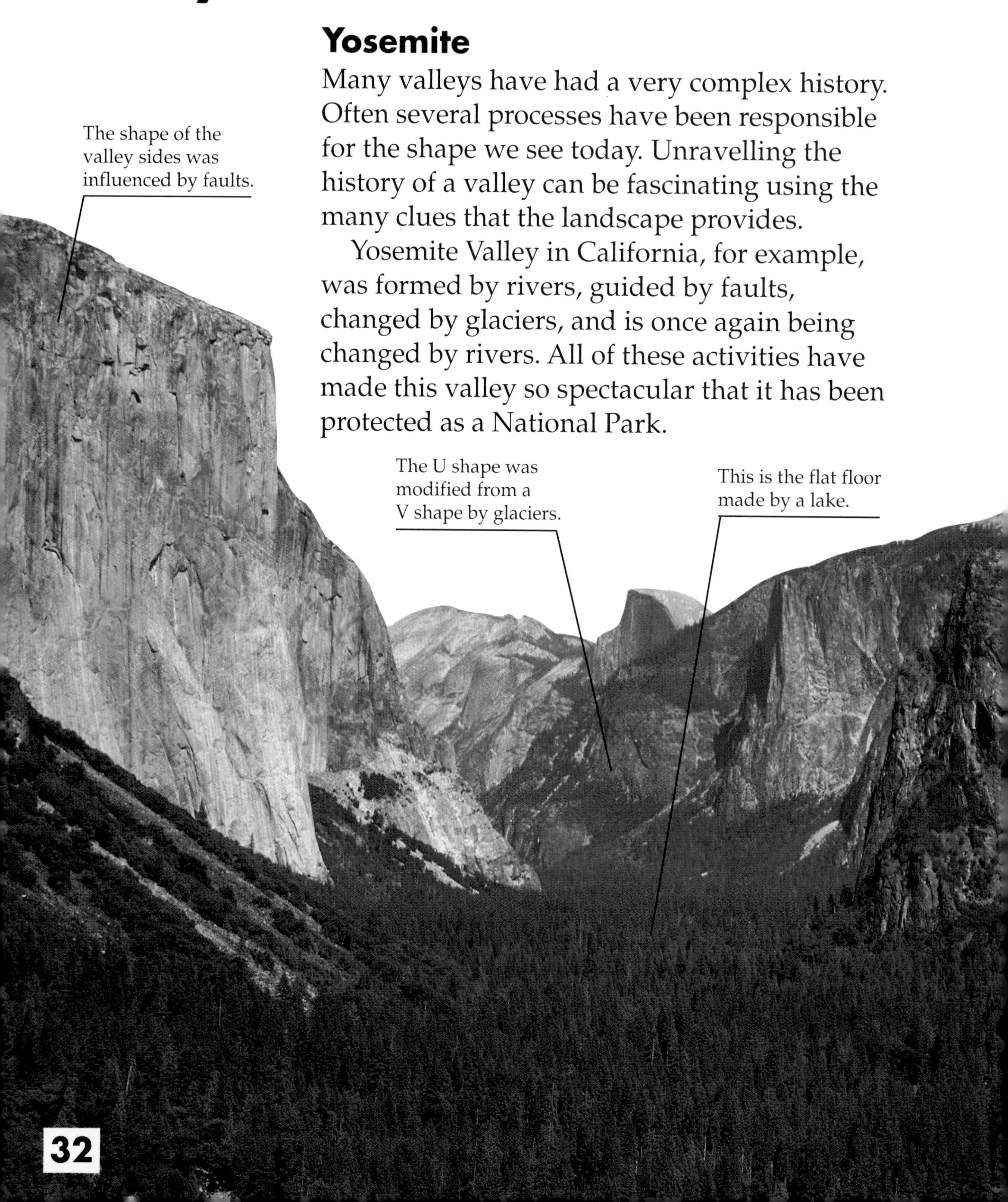

Many of the spectacular shapes of Yosemite have been affected by ancient earthquakes that cracked the crust. The earthquake line, or faults, make the rocks weak and easier to remove.

The great cliff shown in the picture on the right is called El Capitan. It was carved by frost and ice, but the lines of weaknesses decided what shape the ice would cut.

These slopes are being shaped by landslides and rockfalls.

This is a hanging valley.

Rivers are now cutting back into the valley.

A long history

Like many mountain valleys, Yosemite valley is very different today from how it started. Once it was a deep V-shaped river valley with a narrow floor. Then the Ice Age came and glaciers filled the valley, straightening the path and widening the floor. This has been responsible for the sheer walls you can see.

After the ice melted away the floor of the valley became a lake, and large amounts of silt and clay settled on its bed.

Eventually the lake was filled in by the silt and clay. The flat floor of the valley is the old lake floor.

The present river will need a long time to make its mark on this dramatic landscape, but eventually it will cut a gorge into the valley bottom.

The East African Rift Valley

The biggest rift valley in the world is the great East African Rift Valley. Nearly 2400 miles long and up to 30 miles wide, it contains huge lakes.

The Rift Valley begins in the Jordan valley in the Middle East and stretches through the Red Sea and into East Africa, finally reaching the coast in Mozambique. It is one of the world's largest landshapes and it is still growing wider.

The cause of rift valley lakes

Rift valleys commonly contain lakes. As the floor was not formed by a river system, but is the result of Earth movement, the floor of a rift valley does not slope evenly. In fact it may vary quite a lot. Lakes form in the dips.

The biggest Rift Valley lake is Lake Victoria. It is roughly circular and covers an area of 40,000 square miles. Neighbouring Lake Tanganyika (20,000 square miles) and Lake Malawi (19,000 square miles) are long thin lakes.

This picture shows the edge of the East African Rift Valley in Kenya. The edge appears as a low, flat crest. It is, however, many thousands of miles long.

This picture shows a view of the Dead Sea and the Jordan Valley at the northern end of the Rift Valley. Notice how the river, valley and sea each pick out the straight fault line.

The East African Rift valley grows wider by about $\frac{1}{20}$ th of an inch a year. Doesn't sound like much? People near to growing rift valleys think so, because each time the rift moves they feel it as an earthquake!

New words

crust
the outer part of the Earth, containing rocks which are cold and brittle and which will crack when pulled or pushed. The crust is broken into a number of very large slabs, known as plates. The continents are the part of the plates that show above sea level

divide
the line that separates one valley from its neighbor. It is found by looking for the highest points on the ridges of hills between valleys

energy
the ability of a river to move water and carry stones. Energy comes from the amount of water flowing and the speed of the water. The rivers with the largest energy – and those which can cut swiftly into the landscape – are large and fast-flowing

flood plain
the area in the bottom of a valley that is covered with water during a flood. Flood plains are made of material deposited by rivers

glacier
a river of ice that flows from an ice-filled hollow or an ice cap high in mountains

Ice Age
the time, over the last million years, when huge ice sheets spread from the poles. So much water was locked up as ice that the sea levels throughout the world fell by 240 ft and glaciers were able to cut valleys into what is now the sea bed. When the Ice Age ended the ice melted and the sea level rose again, flooding these ice-cut valleys and making, for example, fjords

landslide
a rapid movement of a large slab of soil and rock on a valley side

meander
the pattern of curves that are found on a flood plain. River meanders change size and shape all the time as rivers cut into the outside curve and drop material on the inside

mudflow
a rapid movement of sodden soil or soft clay rock on a valley side

oxbow lake
a small curved lake that shows where a meander used to be. Oxbows are formed when these meander curves become so pronounced that they cut a complete meander loop off

profile
a cross-section of a valley. This is the view you would get while looking upstream. Profiles simply show the shape of the land from the divides to the river

rift
a large trench in the Earth's crust which formed when the crust pulled apart. Rift valleys have steep sided straight valleys and flat floors

soil creep
the imperceptibly slow movement of soil down a slope. Soil creep is the most important soil-moving process on soil-covered slopes

tributary
a river or stream that feeds into a bigger river

Index